Reactive Power Management

Rafael Barreto

CONTENT

1. Basic concepts 5
2. Resistance and reactance 9
3. Effect of a capacitor in an AC feeder 14
    3.1 Active and reactive components of the current ...20
    3.2 Voltage drop 19
    3.3 Losses 21
    3.4 Power factor 22
4. Capacitors and load on the AC feeder 23
    4.1 Voltage rise 25
    4.2 Use of capacitors to manage load 27
5. Devices that demand reactive power 29
    5.1 Induction devices
        5.1.1 Reactive power in the induction motor 33
        5.1.2 Reactive power in transformers 38
        5.1.3 Relationship between primary and secondary current in a transformer 35

6. Compensation of reactive power demanded by motors

6.1 Collective compensation of reactive power 38

6.2 Individual compensation of reactive power 42

7. Capacitors on the high voltage feeder

    7.1 Capacitor banks on distribution feeders 43

    7.2 Capacitor banks at the substation 49

    7.3 The transmission lines as a capacitor 50

    7.4 Influence of transformer taps on load flow 50

8. Reactive power measurement 51

FOREWORD

The time and framework of engineering studies is limited for obvious reasons, only the basics and theoretical aspects are covered in the plan of any engineering study. Details and practical aspects in the different fields are acquired in the praxis and on the go as the different problems present themselves in the life of the recently graduated engineer.

Reactive Power Management deals, in a nutshell, with the different aspects of generation, demand and control of reactive power in the feeders that carry electric energy to the customers who require this energy to do the kind of work needed in the specific productive area.

Reactive power is an undesirable but, at the same time, a necessary element required to use electric energy in the modern world. It is required by all induction devices, like motors, transformers, ballasts needed by the neon light and other types of lighting to produce the magnetic fields that keep them working for us without producing useful work.

Though reactive power cannot be totally eliminated, it can be reduced and controlled by different means in order to make the power delivery more efficient for customers and electric utilities that supply the electric energy. We are sure that this material will be helpful for students and newly graduated electrical engineers from Universities, Technological Institutes and everybody that needs to get informed about the use and control of reactive power.

1.- BASIC CONCEPTS.

What is reactive power and why it has to me controlled?

Reactive power is the power demanded by inductive devices to maintain the necessary magnetic fields for their function in an alternating current circuit. The output of a commercial generator is a sinus wave that varies 60 times per second in US electric system.

The times the sinus wave varies in the unit of time is called the frequency of the wave and is given in Hertz. In the United States is also usual to give electric frequency in cycles per second. In other countries the frequency at which electric current is generated is 50 Hertz or 50 cycles per second. Other frequencies are not usual, but may be found in certain electric systems.

Our analysis assumes that the voltage output is a sinus wave. If the voltage wave loses its sinus shape, many relationships used in the field of electricity are not valid and other methods of analysis are required, for example, harmonics. In all our discussion we will assume that we are dealing with sine voltage and current waves, which is applicable in the vast majority of cases we have to deal with in commercial distribution of electric energy.

In order to make this material practical we will start at by clarifying the basic concepts of alternating current. We will go only as deep as we consider necessary to make clear active and reactive power concepts in an alternating current circuit. In fact, the generator at the power plant generates induced alternating voltage that makes alternating current flow when load is applied to the distribution network.

We may find electric systems that distribute direct current (DC) generated by DC generators, rectifiers or banks of batteries for some specific use. We will use the term alternating current (AC) network to name the alternating current where alternating voltage is generated and alternating current flows and direct current (DC) network to name the network where direct current is generated and direct current flows.

To reach our homes the alternating current must flow through conductors that may be straight or coiled. The conductors of the grid or network are straight and the conductors inside the generator, transformer or motor are coiled. Both the straight and the coiled conductors show and opposition to the flow of the alternating current due to Lenz law. According to this law, the current or voltage cannot change it value instantly in an electric circuit, therefore, we experience like an opposition to the flow of the alternating current everywhere in the network. This opposition is called reactance.

The opposition to the flow of alternating current in the wire, be it straight or coiled, is called inductive reactance In the same way, the opposition to the flow of alternating current in a condenser or capacitor is called capacitive reactance. The term condenser is rather used in electronics, so we will refer to condensers used in the power network as capacitors.

The role of capacitors in the power network is different from those used in electronics. The capacitors used in the power network are manufactured for higher voltage and current. Capacitors may be used at the industry or on the primary feeders of the electric utility that transmits and distributes the electric energy to the customers use. These capacitors are bulky and must be isolated for the voltage on which they are going to be connected. Some attention will be given later to the roll that the high tension capacitor plays in the transmission and the distribution network.

In order to do a better analysis of alternating current behavior, the sine wave is represented by a vector that is an element that has an absolute or modular value and a position in space. Take, for example, force acting on a table. If

the force is applied horizontally, the table will move along the floor depending on the friction. If the same force is applied to the table vertically, the table will be lifted from the floor. The magnitude of the force in pounds or kilograms will be the modular or absolute value of the force, the direction of the movement of the table depends of the angle at which the force is applied, so a vector is defined by a modular value and an angle, both magnitudes are tied together.

Fig 1.1a shows the effect of a force acting parallel, Fig 1.1b shows the same force acting perpendicular to the table. In the first case the table slides along the floor, in the second case the table is lifted upwards. Not only do we have to give the magnitude of the force, but in which direction it is acting.

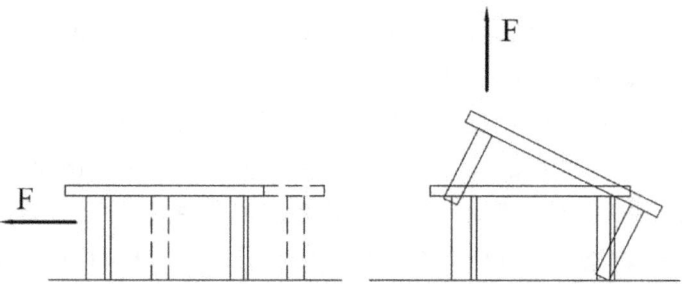

Fig 1.1a - 1.1b Effect of a force F acting in different directions upon a table

In alternating current voltage and current need a modular value given in volt or ampere and an angle to define their relative position.

Figure 1.2 shows how the values on the sinus wave can be transferred to a circle. The V1 value of the voltage can be transposed or projected to the right side on the circle. Note that the modular value of the voltage vector is the same, while the angle varies as we go along on the sinus wave. On the wave the voltage value is the distance from the wave to the X axis, so is it on the trigonometric circle. The effect of this is like the vector was rotating counterclockwise on the circle. If we add a second voltage or current wave, then we will have two vectors rotating together but keeping an angle between them, depending of the relative position of the two sine waves.

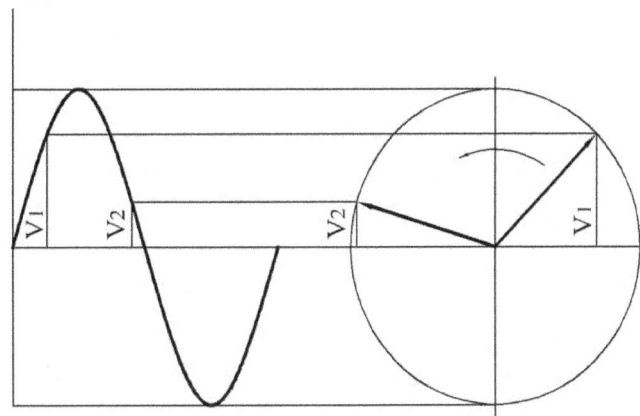

Fig. 1.2 Graphic representation of a sine curve on a trigonometric circle

2. - RESISTANCE AND REACTANCE

We mentioned the term reactance (inductive and capacitive) in the alternating current network. In addition to the reactance we have the resistance of the conductors through which the alternating current flows. Now that we have clarified how the AC waves can be represented as vectors, we will discuss how resistance, inductive reactance and capacitive reactance are related in an AC system.

Due to the impossibility of instant change of current of voltage in any circuit, there will like a delay or lag in the current flowing in the circuit. The opposition to the flow of alternating current on a wire, be it straight or coiled, introduces the inductive reactance, the opposition of a capacitor to the flow of alternating current introduces capacitive reactance on the AC system. The combination of reactance and resistance composes impedance that includes the effect of the reactance and the resistance of the conductor together. Impedance in an AC system IS NOT the arithmetic addition, but the vector addition of resistance and reactance, inductive or capacitive.

The reactive and the resistive component will be placed 900 forming a triangle. The sides of the 900 triangle will be the resistance and the reactance of the wire and the side closing the two components or hypotenuse will be the impedance of the wire or coil.

The resistance will be represented as R; the inductive reactance is represented as XL and the capacitive reactance as XC.

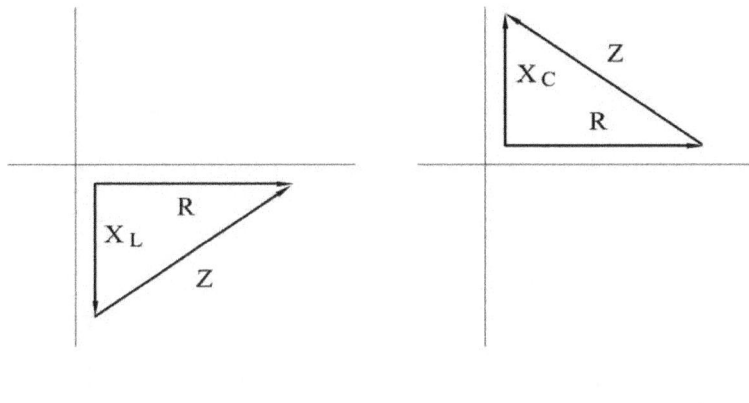

Fig 2.1a   Fig. 2.1b

Fig 2.1a-2.1b Vector relationship between inductive reactance, capacitive reactance and resistance

The capacitive reactance opposes the inductive reactance as shown in fig.2.1a and 2.1b both vectors point in opposite direction.

$$Z^2 = R^2 + X_L^2$$

$$Z^2 = R^2 + X_C^2$$

The capacitor alone is like a load to the system, however, combined in a circuit that has reactance has the effect of modifying the current flowing in the circuit, as we will see later. How can a current that varies 60 times per second produce some work?

Suppose you have a direct current source and a resistor is connected to the circuit. The amount of generated heat will be proportional with the square of the current flowing in the circuit and how long it works.

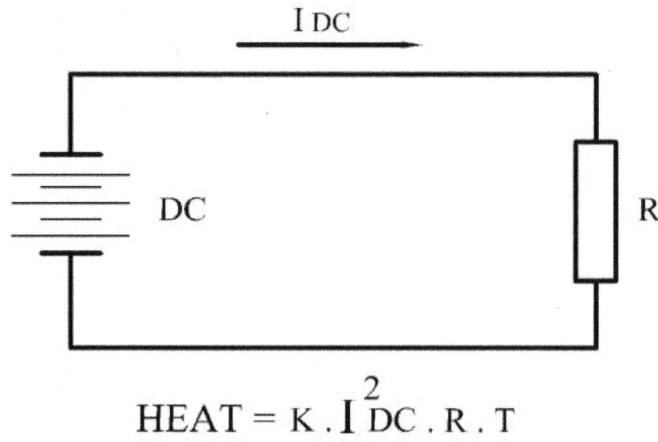

$$HEAT = K \cdot I_{DC}^2 \cdot R \cdot T$$

Fig 2.3 Heat dissipated on a resistor by DC current

Now let's make the current of an alternating source flow through the same resistor

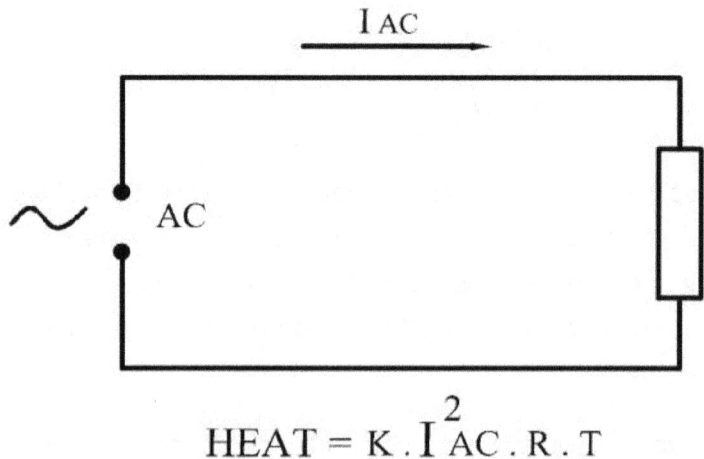

$$HEAT = K \cdot I_{AC}^2 \cdot R \cdot T$$

Fig. 2.4 Heat dissipated on a resistor by AC current

When the amount of heat generated by the alternating current equals the amount of heat generated by the direct current, we say that the direct current value equals the effective value of the alternating current.

As the current varies, so does the amount of heat generated. In order to get the real value of the generated heat, we have to integrate the square value of the current as the wave varies. This way we get the following relationship:

$$I_{eff}^2 \cdot R \cdot T = \sum I_1^2 \cdot R \cdot \Delta t_1 + I_2^2 \cdot \Delta t_2 + \ldots I_n^2 \cdot 2 \cdot R \cdot \Delta t_n \quad (2.1)$$

Taking even smaller time intervals we have:

$$I^2 \cdot R \cdot T = \sum I_1^2 \cdot R \cdot dt_1 + I_2^2 \cdot R \cdot dt_2 + \ldots I_n^2 \cdot R \cdot dt_n \quad (2.2)$$

Integrating this relationship we get:

$$I_{eff}^2 \cdot R \cdot T = R \cdot \int f(I)^2 \cdot dt \quad (2.3)$$

$$I_{eff} = \sqrt{(1/T) \int f(I)2 \cdot dt} \quad (2.4)$$

This relationship represents the effective value of the wave, regardless of its form, and is also called the root mean square of the wave. A multimeter that is able to read voltage or current root mean square is able to show the effective value of the alternating current or voltage it is reading, whatever the wave shape is.

In the same way, if we make a motor run in a direct current (DC) source and another motor run in an alternating current (AC) source, when both motors move the same load and perform the same work we can say that the direct current equals the effective value of the alternating current.

3.- EFFECT OF CAPACITORS IN AN AC CIRCUIT

Due to the fact that the reactance, inductive or capacitive, makes opposition to the flow of alternating current there will like a difference in angle between the voltage and the current wave in the circuit. If the current is flowing through a reactive device, the current will rotate behind the voltage when the two waves are represented by two rotating vectors. Although the two vectors are rotating, the angle between them is the same as long as the inductance-

resistance relationship is the same. In fig. 1.1 we saw how the sine wave can be represented by a vector.

As the voltage varies in the wave, so does it in the circle. The effect is like the voltage vector was rotating counter clockwise. The distance from any point on circle to the X axis will represent the value the voltage takes as the vector rotates.

If we add a current wave and proceed the same way, we will we get another vector for the current that can also be represented on the circle and behaves the same way. As both vectors rotate, the angle between current and voltage will remain the same as long as the conditions of the circuit and the position of the two waves remains unchanged.

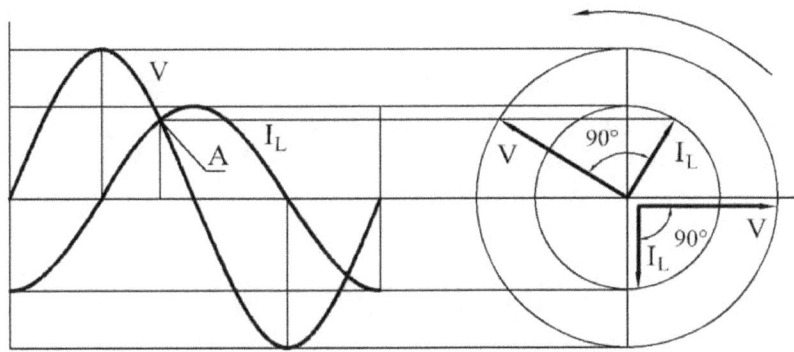

Fig. 3.1 Vector representation of a voltage and inductive current wave

Fig 3.1 shows a voltage and a current wave in the same graphic. Note that the current wave starts when the voltage is already at is maximum this way the current vector rotates behind the voltage vector. As the value of voltage and current at point A is transferred to the circle, we get a 900 angle difference between voltage and current vector with current lagging behind the voltage vector.

Note that when current is growing, voltage is going down towards zero when they reach point A, so voltage vector will be ahead of the current vector.

If the circuit is inductive, the current will be like delayed and will rotate behind the voltage vector. If the circuit is capacitive, the current will try to go ahead of the voltage and the current vector will rotate in front of the voltage vector, as shown in figure 3.2.

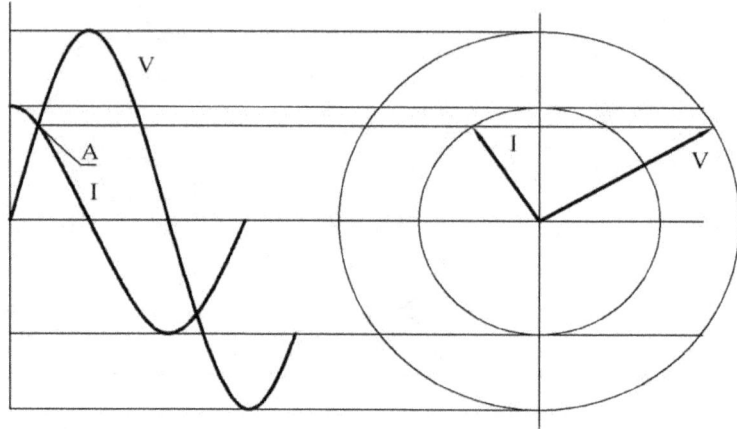

Fig. 3.2 Vector representation of voltage and capacitive current in trigonometric circle

Note that when voltage is zero, the current already has its highest value. Voltage will reach its highest value when capacitive current has already reached zero, this makes the current and voltage vector rotate placing current vector ahead of voltage.

In case of inductive or/and capacitive current the angle between the vectors is the same as long as the waves develop in the same position. If during rotation the angles will be same, it is valid to stop the rotating vectors to analyze their interaction in a vector diagram.

As the capacitive current starts first and the inductive current is delayed the same angle, the capacitive current opposes the inductive current as shown in figure 3.3. The two current waves oppose each other, so the two vectors oppose each other too.

As the vector diagram is a reflection of the wave diagram, we can draw the capacitive current vector in opposite direction to the inductive current vector on the circle as shown in figure 3.3.

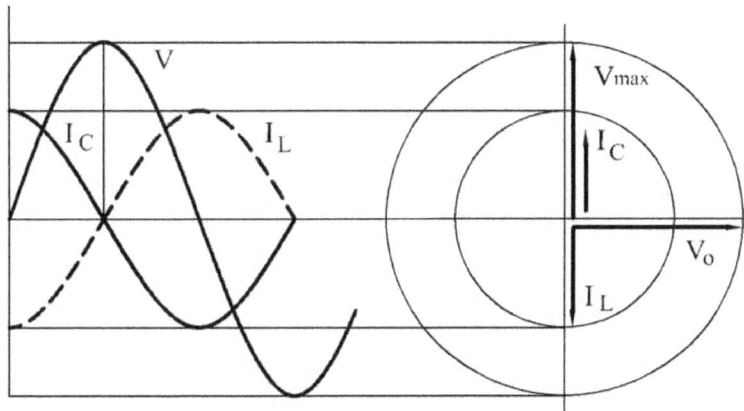

Fig 3.3 Graphic representation of voltage, inductive and capacitive current

Note that when voltage is zero, inductive current vector will placed at 90^0 behind the voltage vector and capacitive current will be placed 90^0 before

the voltage vector. When voltage is zero, capacitive current vector will at its highest point, that is to say 900 ahead of the voltage vector. As inductive and capacitive current will be in opposition, if the capacitive component equals the inductive component, resultant reactive component of the current will be zero. The remaining component will be the effective component of the current IW.

This is specifically the role that the capacitor plays in the alternating current system: reduce the inductive component of the current, thus reducing the overall amount of current flowing and reducing the load in the conductors. This will result in lower voltage drop and lower losses on the feeders that supply the electric energy to the customers.

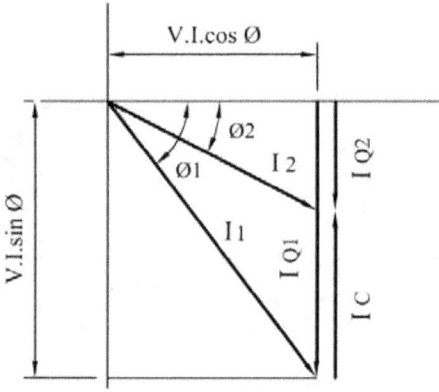

3.1.1 Active and reactive, and capacitive components of the AC current

So far we have been talking about the reactive component of the current, now we have to say a few words about the active component of the current. Figure 3.1.1 shows the components of the alternating current, the active (I_W), the inductive (I_Q) and the capacitive (I_C) components of the current. The active component IW is determined by the load and we have little or no influence on this component. The inductive component of the current can be modified introducing capacitive current in the system. The total current (I_1) will be vector addition of I_W, I_Q and I_C. If capacitive current IC is introduced, original current I_1 will be reduced to I_2.

The active component IW (V.I.cosφ) is the component that performs effective work. This is the component that the energy meter reads, however, the electric system must be able to deliver the whole apparent current I1 demanded by the customer. Voltage drop and losses will be determined by the total or apparent current, that's the reason why the supplier company shows interest in reducing the total or apparent current to a reasonable level. In figure 3.1.1 the introduction of capacitive current IC reduces the apparent current from I1 to I2. Reduction of the total current reduces voltage drop,

losses and releases capacity that can be used to add additional load with little or no investment.

3.2 VOLTAGE DROP

In the AC system the angle between current and voltage introduces an active and a reactive component in the current. The voltage drop will have therefore two components: one resistive, lying alongside the current and the other inductive is placed 900 to the current vector I. As the resistive component of the voltage drop points in the same direction as the current it is said that they are in phase..

Figure 3.2.1 shows the approximate voltage drop components in an AC circuit: $V_1 . I_W$ and $V_1 . I_L$

$$V_1 . IW = V_1 . I . \cos\phi \qquad (3.1)$$

$$V_1 . IL = V_1 . I . \text{sen}\phi \qquad (3.2)$$

V_1 is the voltage before the load is connected to the feeder and V_2 is the voltage available after substracting the voltage drop.

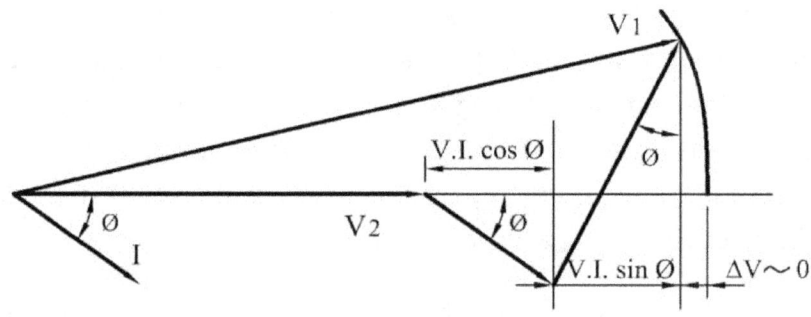

Fig 3.2.1 Components of voltage drop

Taking $\Delta V = 0$ is good enough approximation for engineering purposes, so we can say that a good approximation to the voltage drop in an AC feeder will be

$$\text{Voltage drop} = V_1 . I . \cos\phi + V_1 . I . x . \sin\phi \qquad (3.3)$$

Note in figure 3.2.1 that there will be an angle between the voltage V_1 at the beginning of the feeder and voltage V_2 at the end, however, voltage drop refers only to modular difference between initial and final voltage value. What is important is the modular difference that would be registered by a voltmeter between V_1 and V_2.

3.3.- LOSSES

According to Ohm Law, power losses equal squared current value times resistance, in AC and DC feeder as well. If power is multiplied by the time it is demanded, we get energy.

$$\text{Loss (watt)} = I^2 \cdot R \qquad (3.4)$$

The magnitude of the flowing DC current can be varied only varying the load, however, in the AC feeder the flowing current can be influenced by the amount of reactive current in the feeder. Refering to (3.4)

$$\text{Loss}_1 = I_1^2 \cdot R \qquad (3.5)$$

$$\text{Loss}_2 = I_2^2 \cdot R \qquad (3.6)$$

Due to the quadratic relationship if we make current

$$I_2 = (1/2)I_1 \text{ 1 then}$$

$$\text{Loss}_2/\text{Loss}_1 = (I_1/2)^2 \cdot R/(I_2^2) \cdot R = \tfrac{1}{4} \qquad (3.7)$$

It means that by reducing the current flowing on the feeder to a half, we can reduce losses to a fourth of the original value.

3.4 POWER FACTOR

Power factor is defined as the cosine of the angle between current and voltage on an AC feeder.

Referring to figure (3.1.1) note that IW = I. cosϕ is the *effective* or *active* component of the current. This is the current component that performs work and is *proportional to the energy registered by the watt-hour meter*, regardless of angle between current and voltage. The purpose of introducing capacitive current to reduce reactive component, thus total current, is solely to reduce voltage drop and losses on the feeder, active power and energy remains the same, it is determined by the active load.

Improving power factor means reducing the angle between current and voltage and bringing the total current that must be supplied by the source as close to the active component as possible.

The product of voltage and active current will be the *active*, the product of voltage and reactive current will be *reactive*, and the product of voltage and total current will be the *apparent* power taken by the load. As you may notice, the load that the source or utility company must supply is the total or apparent power demanded by the load taken by the customers on the feeder.

The apparent power unit is usually the *volt-ampere* (VA), the effective power usual unit is the watt (W), the usual unit for the inductive reactive power is the *reactive volt-ampere* (VAR), the usual unit for the capacitive reactive power is the *capacitive volt-ampere* (CVA)

The cosine of $0°$ is 1.0, this is the highest possible power factor on a feeder. Demanding 50 A active current at power factor 0.8 means that the source must supply 50/0.8 = 62.5 A apparent current, but will generate revenue only for 50 A. This is the component used by the watt-hour meter for energy registration. As higher current introduces higher losses and higher voltage drop on the feeder, some utilities include a clause that introduces a penalty for low power factor on industrial customers where high apparent load is demanded and relative little active power is used to perform work.

Writing losses relationship as a function of power factor we get:

$$Loss_2/Loss_1 = (I_w/\cos\phi_2)/(I_w/\cos\phi_1) \quad (3.6)$$

$$Loss2/Loss1 = (\cos\phi_2/\cos\phi_1)^2 \quad (3.7)$$

Resuming: *The higher the compensated power factor related to the original power factor, the lower the losses.*

4. CAPACITORS AND LOAD ON THE AC FEEDER

The static capacitor is the cheapest and simplest way to inject capacitive reactive current in the system. According to Ohm Law

$$I_C = Voltage/X_C \quad (4.1)$$

Where XC : the reactance of the capacitor in ohm.

Figure 4.1 shows the effect of introduction of capacitive current on the feeder.

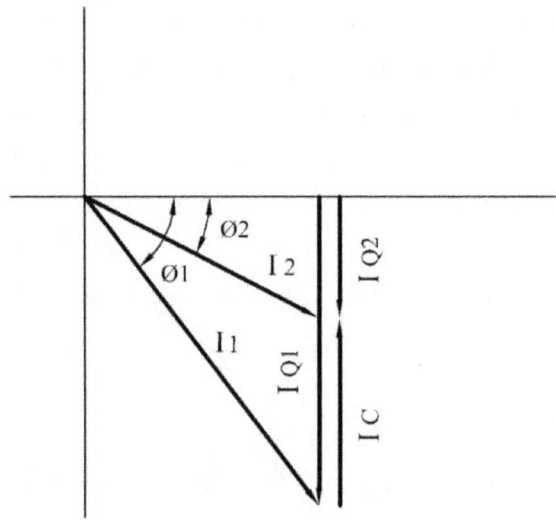

Fig 4.1 Reduction of the load current I1 Introducing current IC

Note that introducing IC reduces the inductive current component from IQ1 to IQ2. As a result, the total apparent current will be reduced from I1 to I2 in figure 4.1

The capacity of the condensers used in electronics is normally given in Farad, microfarad, nanoFarad, etc. The capacity of the condensers (capacitors) used for power factor correction is given in volt-ampere capacitive, or kilovolt-ampere capacitive (CkVA). This is the unit used in the power field when referring to capacitors in the US.

The capacitive current is calculated usually using the following expression.

$$I_c = CkVA/kV \qquad (4.2)$$

This expression is good for one-phase values. For three-phase values we have to write:

$$I_c = (CkVA/\sqrt{3})/kV \qquad (4.3)$$

4.1 Voltage rise

Returning to relationship (3.1) we can write the way the capacitor reduces voltage drop

$$\Delta V = R \cdot I_1 \cdot \cos\phi + XL \cdot (I \cdot \text{sen}\phi - I_c) \qquad (4.4)$$

$$\Delta V = R \cdot I_1 \cdot \cos\phi + I_1 \cdot XL \, \text{sen}\phi - X_L \cdot I_c \qquad (4.5)$$

So, based on this relationship we can say that the voltage rise introduced by the capacitor will be proportional to the product of the capacitive current and the reactance of the feeder, whether there is reactive load or not.

Sometimes it is advisable to connect a choke in series with the capacitor to increase the total reactance through which the capacitive current must flow ($X = X_L + X_{CH}$). In this case a resonance study must be performed to avoid major problems.

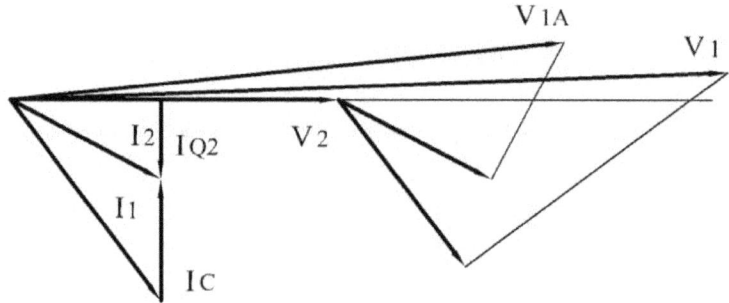

Fig 4.1.1 Voltage drop modified by capacitive current

Figure 4.1.1 shows the voltage drop vector diagram modified by the presence of the capacitive current IC.

Note that the presence of capacitive current I_C makes the modular difference of $V_1 - V_{1A}$ smaller in comparison with the difference $V_1 - V_2$ when there was no capacitive current in the system. It means the voltage drop is smaller after the capacitor is introduced in the system. The components on the vector diagram have been exaggerated for better understanding.

The term $-X_L \cdot I_C$ in relationship 4.4 leads to a very interesting fact. As the voltage rise equals the product of the capacitive current IC and the inductive reactance XL and has nothing to do with load, it is constant on the feeder all the way to the source. Voltage rise will be present regardless of whether we have any load on the feeder or not, it is always present.

Fig 4.1.2 shows the influence of the capacitor in the voltage level along the feeder. The dotted line represents the original voltage level before capacitor connection. Note that the voltage level decreases toward the end of the feeder and the capacitor introduces a voltage rise from the point B where it was connected towards the source represented by point A.

If the load distribution beyond point B to the end is the same, voltage profile will have the same slope through the end of the feeder.

Note in figure 4.1.2 that voltage rise $I_C \cdot X_L$ is always present, even when load is zero. The farther away is the capacitor, the greater will be the reactance to the source and the greater will the voltage rise.

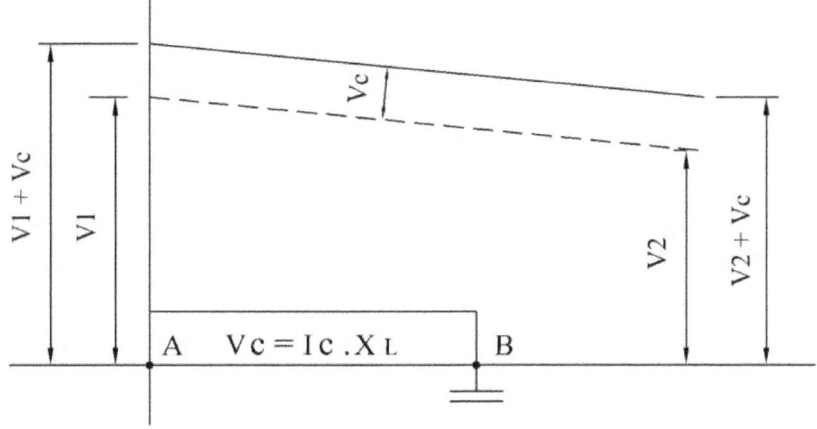

Fig 4.1.2 Voltage rise introduced by a capacitor on a feeder

4.2 Use of capacitor to manage load

Reducing the current on the feeder not only has the advantage or reducing voltage drop and losses. Current reduction releases capacity on the feeder that can be used to add more load without increasing capacity.

Fig 4.2.1 shows the vector components of active P (kW), inductive Q_L (kVAR) and apparent S (kVA) power in a given situation.

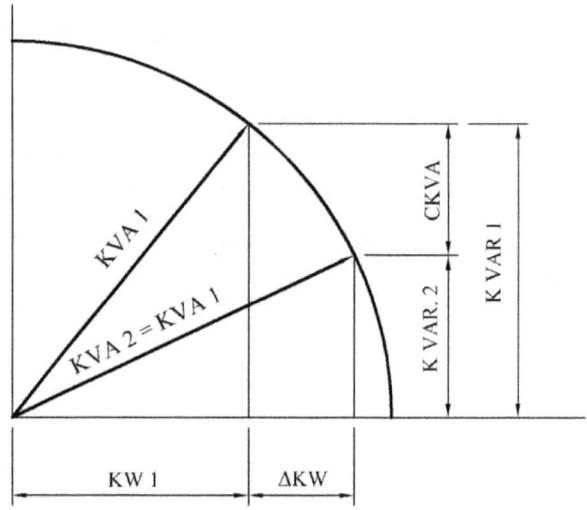

Fig. 4.2.1 Adding extra capacity using capacitors

There is a need to add extra load represented by ΔP at the same power factor of the existing load. In that case the situation can be solved by adding QC capacitive power that will reduce the existing reactive power from kVAR$_1$ to kVAR$_2$, this way we keep total load at the same value before and after new load addition (S1 = S2) and we will have extra load ΔP available. As long as we keep vector S on the circle, the load remains constant.

Let's write the relationships down in a mathematical relationship:

S1 = S2

$$(P_1 + \Delta P)/\cos\phi_2 = P_1/\cos\phi_1 \quad (4.5)$$

$$\cos\phi_2 = ((P_1 + \Delta P)/P_1) \cdot \cos\phi_1 \quad (4.6)$$

$$\cos\phi_2 = (1 + \Delta P/P_1) \cdot \cos\phi_1 \quad (4.7)$$

This will be the power factor required to release ΔP extra capacity

Capacitors installation does not allow any amount of power increase. Observing fig. 4.2.1 note that the maximum extra load that can be added is when $\cos\phi = 1.0$. Re-arranging (4.7)

$$1 = (1 + \Delta P/P_1) \cdot \cos\phi_1$$

$$\Delta P = (1/\cos\phi_1 - 1) \cdot P_1 \quad (4.8)$$

If $\cos\phi = 1.0$, then $\Delta P = 0$, no extra capacity can be added.

The use of capacitors for capacity increase should be considered when capacity transformer and/or feeder capacity increase is unfavorable. Decision must include economic considerations also.

5. DEVICES THAT DEMAND REACTIVE POWER

5.1 Induction devices

Induction motors, generators and transformers are of vital

importance to generate, transmit and distribute electric energy. They all function on the principle of interaction between current flowing on a conductor and a magnetic field.

We saw that the relative movement of a conductor perpendicular to a magnet field induces a voltage on the conductor. If the conductor is closed in a loop, electric current flows and this flow of current has a magnetic field associated with it. This is the corner stone of the use of electricity. We say relative movement because the same effect is attained if the conductor is moved in the magnetic field, or the magnetic field is moved over the conductor.

The more turns the loop has, the greater the induced voltage, so induced voltage depends on the number of turns. The more intense the magnetic field, the greater the induced voltage, so the induced voltage depends on the field intensity. The greater relative movement of the magnetic field across the conductors, greater induced voltage is obtained. We can say then that the induced voltage depends on the frequency of the magnetic field associated to the current.

AC devices, like generators, motors and transformers work on the principle of a magnetic field moving through conductors. In the case of generators and motors the magnetic field rotates around a round core called the stator. In the generator the rotor or central part contains a coil through which flows a DC current that creates a magnetic field. This magnetic field sweeps the wires wound on the stator and voltage is induced.

In the case of the induction motor, the rotating field on the stator induces current that flows in a sort of short circuited bars in the rotor like those in which the hamster runs. These devices are called squirrel cage rotor due to the likeness to the cage in which the hamster and/or the squirrel runs to entertain us.

In the generator the coils are wound at 1200 to one another. The magnetic field on the rotor will sweep over the coils cyclically, therefore the induced voltage will be in three waves that will also be displaced 1200 represented by three rotating vectors on the circle displaced 1200 between them .

In the motor the coils are wound displaced 1200 too. The three-phase voltage imposed cyclically on the coils of the motor will make the effect of a rotating magnetic field around the core of the motor. This rotating magnetic field sweeps the short circuited bars on the rotor and induces current that interacts with the rotating magnetic field in the core. As a result of this interaction a force is created that makes the rotor move.

As the rotor of the motor rotates it is more convenient to consider the torque associated to this force to characterize the motor.

5.1.1 Reactive power demanded by inductions motors.

In order to keep the rotor rotating and produce its torque, the induction motor demands reactive power to keep its magnetic field and move the load attached to its shaft. This demand is determined by the nameplate power of the motor and the torque it is expected to give. The power delivered by the induction motor is given mainly in kilowatt, but the term horse power is also used.

The magnetic lines jump from the stator to the rotor through the air in the gap between the rotor and the stator. On motors of small capacity this gap is relatively large in comparison with the size of the motor. The more magnetic lines closing through the relatively larger gap, the larger the reactive power demand and the lower the power factor.

On larger capacity motors the gap between stator and rotor is comparatively smaller, the reactive power demand will be relatively smaller and the power factor will be better than in small motors. In the case of small capacity motors, the power factor stands between 0.7 and 0.75. In the case of larger inductions motors the power factor can be between 0.8 and 0.9. In the case of quite small motors used in household devices, like hair-dryers,

refrigerators, fans, etc. the power factor can be quite low, between 0.6 and 0.7.

Fig 5.1.1 shows the way active and reactive power varies in the induction motor. While the active power drops to practically zero, reactive power demand drops to 75-80% of reactive power demand at full load. The active power demand of the motor at no load will be the losses in the core and the windings and a small power demand originated by friction.

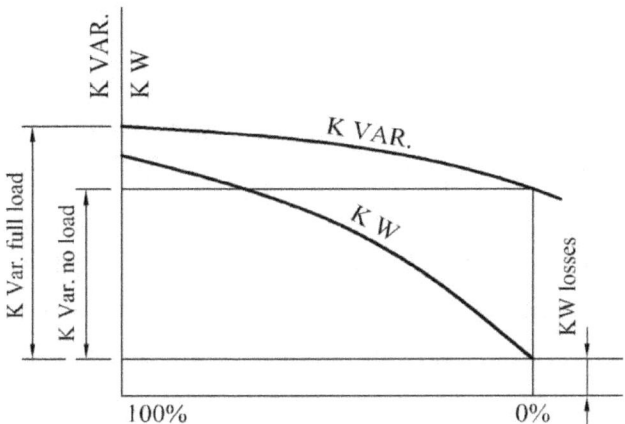

Fig 5.1.1 Active and reactive power in an induction motor depending on load

A reduction in the load of the motor will not reduce reactive power in the same amount as active power, it needs to keep reactive power to keep the rotor moving.

5.1.2 Reactive power demand in transformers

The transformer is composed of one iron core on which two coils are wound: the primary and the secondary. The primary and the secondary nomination is conventional, any coil can be primary or secondary, but usually the primary coil is the one that carries the magnetizing current to the transformer.

With the secondary coil open, the primary coil takes current to make the magnetic field flow along the core and to cover the active losses generated by the currents flowing in the core due to the changing magnetic flux. This current is called *eddy current.*

The sweeping of the magnetic flux over the secondary coils induces voltage on the secondary coil proportional to the number of turns over which the magnetic flux swipes. If the secondary coil has half the number of turns of the primary, the induced voltage will be approximately half of the value of the primary voltage. If the coil with the smaller number of turns takes the current to excite the magnetic flux, then this coil will be the primary coil. So any side of the transformer can be either primary or secondary depending on which side carries the necessary current to make the magnetic flux flow along the core.

The coils are also called the windings of the transformer. The primary winding carries the current demanded by the load plus the magnetizing current, that is why the reactive demand on the primary side will be greater and the power factor lower than that of the load, as shown in figure 5.1.2

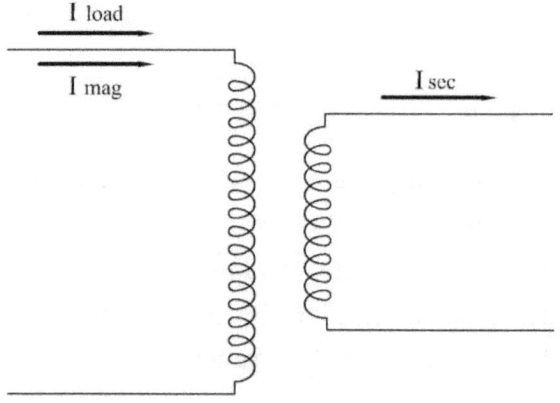

Fig 5.1.2 Graphic representation of a transformer showing primary and secondary load, and magnetizing current

The magnetizing current is also called the no load current. The no load current contains the reactive component necessary to maintain the magnetic flux flowing on the core plus the losses introduced by the eddy current. This current introduces losses in the iron core that equals the square eddy current times the resistance of the core. Generated heat will be proportional to the product of the loss power and the time the transformer is on. Losses can be only modified by the material and the make of the core. Keeping transformer's no load loss low is important, as the transformer is usually on all the time.

In order to make eddy currents lower, the core of the transformer and other rotating AC machines is formed by isolated plates packed and firmly held together to divide the eddy current and limit core losses. If the plates are not firmly held together, a humming sound will be produced by the forces associated to the changing magnetic flux.

There is little or nothing we can do to reduce reactive power demand in the transformer, as it is determined by the necessary magnetic flux to induce nominal or nameplate voltage on the secondary side.

5.1.3. Relationship between primary and secondary current.

The power demanded by the load must be the same on each side of the transformer.

$$P_{prim} = P_{sec}$$

$$V_{prim} \cdot I_{prim} = V_{sewc} \cdot I_{sec}$$

If
$$V_{sec} = 1/2 \cdot V_{pim}$$

then
$$I_{prim} \cdot V_{prim} = I_{sec} \cdot (1/2) \cdot V_{prim} \qquad (5.1)$$

V_{prim} cancel at each side of (5.1), therefpre,

$$I_{prim} = 1/2 \cdot I_{sec} \qquad (5.2)$$

Higher voltage is associated to smaller current and lower voltage to higher current. The primary and secondary current will be in inverse proportion to the primary and secondary voltage.

Dividing the larger number of turns by the smaller number of turns we obtain the ratio of the transformer. The ratio will be proportional to the larger voltage divided by the lower voltage of the transformer. Knowing the ratio of the transformer we can determine current or voltage on either side of the transformer. If the ratio is 10 and the secondary current 100 A, the primary current will be 100/10 = 10 A. If we connect 100 kV on the primary side of the transformer we will get 100/10 = 10 kV on the secondary side.

In order to reduce current and at the same time losses, high power is transmitted at high voltage. Of course, there is an economic limit, as higher voltage requires larger transformers, higher number and taller towers and higher insulation level at higher costs.

6.- Compensation of reactive power demanded by motors.

At start, the rotor of the induction motor works a short circuited transformer producing a large starting or inrush current. Large capacity motors sometimes have to be started at reduced voltage to reduce the starting current to reasonable levels. It is accomplished by so called compensators, these are nothing but autotransformers that feed reduced voltage to the motor

in order to reduce starting current. Sometimes the three phase leads of the motor are separated, so the windings can be connected either in Y or in Δ.

A method to reduce reactive power consumption is to start all the way around : start with the windings connected in Δ to guarantee the starting torque, then switch the coils connection to Y. When starting in triangle the coils will receive full voltage and will be able to give full torque to start the process. Switching to Y will reduce the voltage to $V/\sqrt{3}$. This is accomplished by means of automatic starters that switch the connections of the coils after the motor has reached full torque.

A previous study of the torque demanded by the load is required before applying this method of reactive power reduction to guarantee that the torque will be enough to keep the load moving.

We have to bear in mind that the starting torque is proportional to the squared voltage so the performance of the motor must be studied before deciding how to start and run it.

6.1 Collective compensation of the reactive power.

The most widely spread way of compensation of reactive power demand at an industrial center is by means of the static capacitors. It is important to note, however, that the capacitive current flows from the capacitor to the source, it means that the capacitor will be effective only from the place where it will be installed back towards the source. Inside the inductive devices nothing changes, reactive power will not be reduced inside the

induction machines. Only the feeder will benefit carrying smaller current, having less losses and little higher voltage. If the total current can be lowered at an industrial center by means of static capacitors, the feeder will carry lower load and eventually extra load can be added to the feeder if necessary with little or no extra investment as we discussed in 4.2.

If the load is diversified, capacitor banks are connected at one or more load centers that feed the different inductive loads. The capacity of capacitors is a discreet value, so we have to use the available capacities in industry. That is why we will probably have to add several capacitors of the same capacity to come close to our needs. The capacitors are grouped in banks called capacitor banks.

A previous study of the nature and diversity of the load is needed to determine where and what capacitive load will be installed. The objective of the capacitor may be to reduce load on the feeder reducing voltage drop and losses at the same time. The objective may be to comply with requirements of the electric supplier to keep the power factor at a given level. The supplier may charge a penalty for low power factor or for high reactive demand. Whatever the purpose of the capacitor connection a previous study of the reactive load is required.

The most accurate tool to start the study is a graph of kVAR/time where the reactive demand of the different loads will be reflected and how long each of them is on to determine the reactive power demand at every hour of the day.

The most accurate way of determine the combined reactive load is by means of a device that records on a band of paper the combines reactive load during a period of time. If such device is not available, then the reactive load must be estimated by means of current measurements and estimated power factor. If the motor is fully loaded, then the nameplate power factor can be used. If the motor is not fully loaded, then a power factor lower then nameplate must be taken to estimate the reactive power demand. 75% of full reactive power can be used as an approximation.

The time of the day at which the load is demanded is important to estimate a reactive power demand at each hour during the period of time the customer is working.

Fig 6.1.1 shows what should be the graphic of reactive power demand in case of one shift industrial customer.

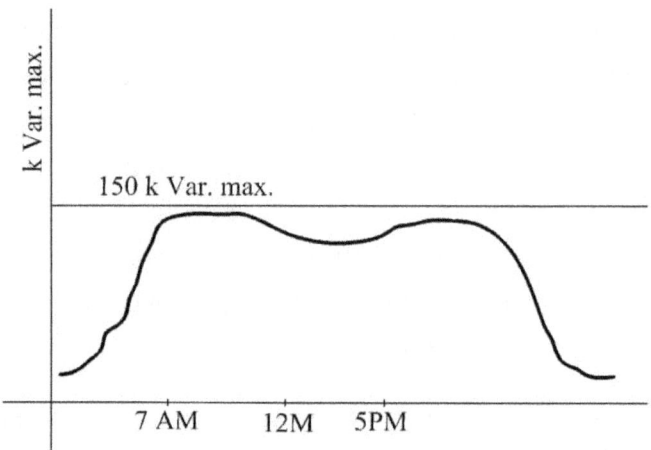

Fig. 6.1.1 Reactive power distribution at a one-shift customer

The load (active and reactive) should start to go up around 7 AM when the shift starts. There can be a decrease of the load due to the lunch break and there is a noticeable reduction around 4 PM when the shift ends. The load after 5 PM is mainly lighting and maybe ventilation and some small refrigerator or vending machine working through the night.

Next day the same cycle starts again. It will be probably not exactly like the bay before due to different usage of the tools and machinery in the shop or factory.

When everybody goes home and the shop stops, the load, including the reactive, will drop to its lowest level after 5 PM. If the capacitor stays connected, then the capacitor itself becomes a load for the feeder keeping the

same level of voltage rise IC.XL. What to do in this case? If the voltage rise introduced by the capacitor is not substantial, we have the choice to leave connected full time.

If the voltage rise is substantial, we have the choice to disconnect the capacitors manually or through a voltage relay that senses the voltage level and sends a signal to disconnect the capacitors when the voltage reaches a given level. The relay can connect the capacitor in the morning when the voltage level start to drop. The capacitor can be controlled by a device that senses the reactive power, or the power factor level and connect it if the power demand or the power factor goes over the limit set on the relay.

The diagram in fig 6.1.2 shows a Load Center that supplies power to different motors that perform different work at different machines. If the machines are small, the choice is to connect the capacitors at the bus of the load center. Note that the reactive load reduction takes place only from point A, where the capacitor is connected, towards the source. Beyond point A towards the load, the reactive load is unchanged. M1 , M2 ... etc. represent the load composed of induction motors. The capacitor bank compensates reactive load demanded by all the motors on the load center.

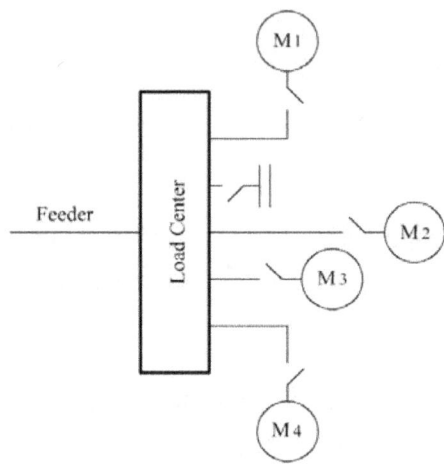

Fig 6.1.2 Schematic load center with centralized reactive load compensation

6.2 Individual compensation of the reactive power.

Individual compensation can be considered in the case of large motors connected all the time or during a considerable number of hours. In this case the reactive power demanded by the motor is taken as a basis to choose the CkVA needed to keep the reactive power of the motor at a minimum. The capacitor will be effective from the point where it is connected towards the source.

If the capacitor is connected direct to the motor as a unit, the overcurrent protection of the motor must be re-calibrated, taking into account the new lower current level. The closer to power factor 1.0, the more capacitors are needed, so the amount of reactive power to compensate is determined by the economic use of the capacitor. When applying individual compensation the no load reactive power demand is used to determine the necessary

CkVA, this will equal 75-80% of the name plate reactive power demand.

Fig 6.2.1 shows schematically the individual compensation of a large induction motor.

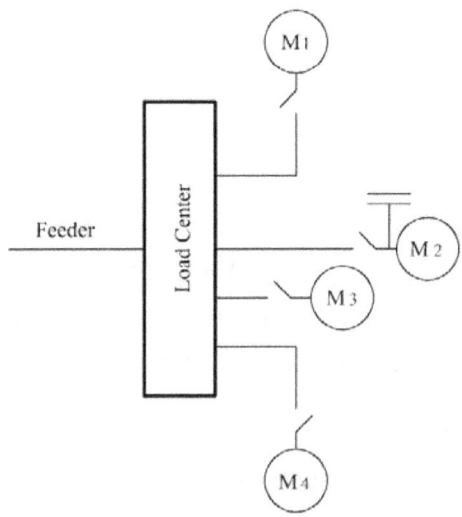

Fig 6.2.1 Schematic individual compensation of reactive power at a motor

7. – CAPACITORS ON THE HIGH VOLTAGE GRID

7.1 Capacitor banks on the distribution feeders

It is the interest of the utility companies that supply energy to have a good voltage level and low losses on their feeder.

In areas where there is high density of household load the electric energy supplying companies prefer bulk reactive load compensation and install capacitors on the high voltage feeder as a practical en less expensive way to keep a decent voltage level.

Urban underground networks are usually interconnected and the capacitive power flow must be analyzed by means of computer programs that take into account the power flow on the underground network. The cables themselves show high capacitance, so a computer program must be run to make the decision about where to install the capacitors, depending on voltage and load distribution at different time along the network. Overhead feeders are more flexible, easier to handle, less costly but less reliable than underground cable networks.

Reactive load is easier to manage than active load, capacitors are also installed on the high voltage overhead feeders. As we saw before, capacitors are installed in banks composed of several units, usually of the same capacity, to make the required CkVA for the load.

The domestic load, active and reactive, is not evenly distributed along the overhead feeder. As we go from the end of the feeder to the substation, load

is being added to the feeder. The feeder load is shown schematically in fig. 7.1

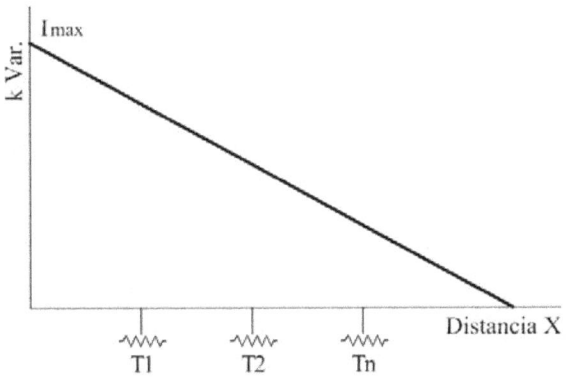

Fig 7.1 Schematic distribution of reactive power along a feeder

The different distribution transformers that supply customers along the feeder add increasing load as we along the feeder to the substation. The graphic in figure 7.1 shows the approximate

reactive load distribution along the feeder to the substation.

As the active load is determined by the customer and we have little or no effect upon this load composed of TV, freezers, air conditioning sets, light, etc. we will deal only with the reactive load that all those household devices demand. The function of this load will be considered as $I_{QMAX} \cdot (1-X)$ where

I_{QMAX} is the highest reactive current at the substation

X is the distance from the substation to the end of the feeder

If higher voltage is all we need, then it is advisable to install the capacitor bank as far away of the substation as possible to maximize reactance XL as the voltage rise will be $I_C \cdot X_L$.

Usually the capacitor bank is installed at some intermediate point on the feeder. There is a point, however, where voltage rise can be combined with minimum losses. Figure 7.2 shows the influence of the capacitor bank on the reactive load of the feeder.

Considering the length of the feeder as 1.0, the total loss after capacitor installation will be the integral of the current distribution from 0 to the point "a", including the capacitive current, plus the integral of the rest of the reactive current from the point "a" to the point 1.0, that is the end of the feeder.

The total loss will be the loss considering the capacitor connected to point a plus the remaining loss without compensation from point a through end represented by point 1.

The reactive current distribution increases as we come closer to the substation and it is in inverse function to the distance, so we have to integrate and writ the following relationship:

$$\int_0^a [I_{Qmax}(1-x) - IC]^2 \cdot R \cdot dx + \int_a^1 [I_{Qmax}(1-x)]^2 \cdot R \cdot dx$$

One integration is from 0 to point a, the second is from point a, where the capacitors are installed, to 1, which is the end of the feeder.

If we want to compute the loss in per unit of the total loss before the capacitor bank was added, we have to divide the result of the addition of the partial losses by the total loss before the installation of the capacitors

$$\int_0^1 [I_{Qmax} \cdot (1-X)]^2 \cdot R \, dx \qquad (7.2)$$

Leaving the computation of the integrals to the reader for a rainy day, we arrive to the following relationship:

$$P_{losses} = 1 + 3 \cdot a \cdot (I_C/I_{QMax}) \cdot [-2 + a + (I_C/I_{Qmax})] \quad (7.3)$$

Relationship 7.3 generates a set of curves that show how losses vary along the feeder as a function of I_C/I_{Qmax} what is the same as CkVA/kVAR. The former is used for illustration purpose, the latter is used for practical purposes. Fig. 7.2 shows the set of curves.

The graphic 7.2 reflects only three curves for I_C/I_Q = 0.2, 0.6 and 0.9

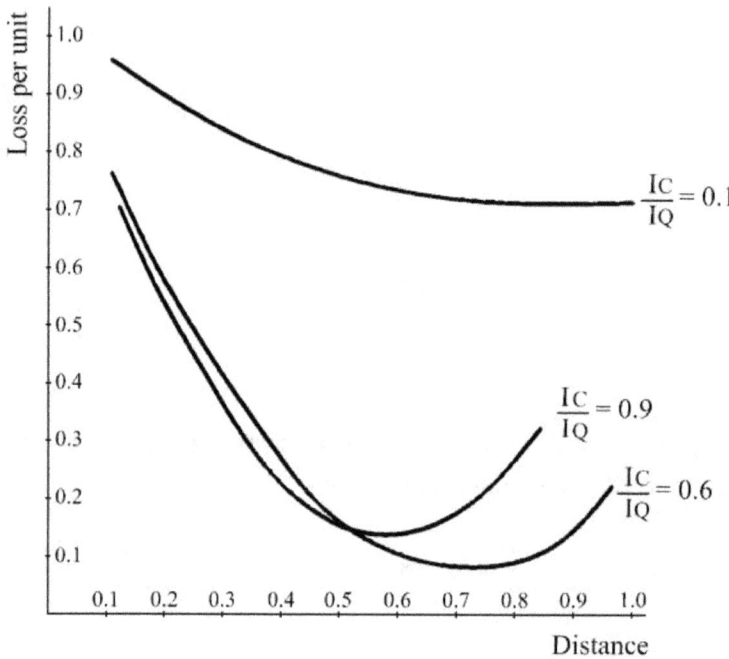

Fig 7.2 Losses per unit as a function of I_C/I_Q placing the capacitor at different points along the feeder

This graphic shows something interesting: there is a definite capacitive load or which the losses related to the reactive component of the current are the smallest, I_C/I_{Qmax} (CkVAR/kVAR) = 0.6 installed at 70% of the distance from the substation.

We can state as a rule of thumb that roughly *the most efficient location for a capacitor bank on a distribution feeder is 2/3 of the maximum reactive load at the substation installed at 2/3 of the distance from the substation.*

Of course, we have to remember that this result is valid for a given condition for the highest reactive load at the substation at a given time.

Another interesting conclusion to be drawn from fig 7.3 is that the smaller the capacitor bank, the farther on the feeder is located the point of maximum performance. Watch curve for $I_Q/I_C = 0.1$. The point of best performance is toward the end of the feeder. Curve for $I_Q/I_C = 0.9$ has a minimum too, but loss reduction is lower than $I_C/I_Q = 0.6$.

Some capacitor banks are controlled by means of voltage relays and are switched off to avoid excessive high voltage during the time of light load after midnight and is switched back on when the load grows.

7.2 Capacitor banks at the substation

The reactive power flows in either direction on the high voltage transmission network. Studies are made on computers about the reactive and active power flow in the high voltage network that cannot be supplied only by the distribution capacitor. For his purpose larger capacitor banks are installed at some substation where capacitive load is required according to the results of the computer software that controls the reactive flow on the transmission system. These large capacitor banks are connected and disconnected at the convenience of the load dispatch center.

7.3 Transmission lines as a capacitor.

Any two metal conductor close to one another makes a capacitor. The high voltage of the transmission lines and the higher cross section of the conductors turn the transmission line into a capacitor. A transmission line of

138 kV 50 miles long can inject about 4 – 6 CMVA into the transmission system. This fact is also taken into account by the power flow software to determine the direction in which active and reactive flow in the system and whether additional capacitive load is needed at the substation.

The capacitive power generated by distribution feeders is negligible, so static capacitor are used to modify the reactive profile of the feeder.

7.4 Influence of transformer taps in load flow.

The distribution transformers and the large transmission transformers have taps that can change the transformer ratio by adding or removing turns to the coil. The taps are usually installed on the high voltage side as the cross section of the primary coil is smaller. Changing the taps of the distribution transformer requires disconnection of the transformer. Adding turns in the primary coils increases the transformer ratio, it results in a lower voltage on the secondary side. Removing turns have the opposite effect. In the case of the large transmission transformer tap changing is performed under load, the transformer doesn't have to be disconnected for this purpose.

Changing the taps of a large transformer will add or remove turns from its winding as well. If the transformer is connected as a link between two sections of the system at different voltage, adding or removing turns will result in a higher or lower voltage at that point. If the voltage is made higher, the interconnection with the system will make a current flow from the point of higher voltage to the point or points where voltage is lower. This current will have a reactive character, as it is not the result of active load demand. As a result, the manipulation of the taps in the large transformers can change the reactive power flow on the system. This effect is sometimes used by the load dispatch to modify the reactive power flow in one direction.

8. REACTIVE POWER MEASUREMENT

The reading of a watt-meter or watt-hour meter is proportional to the product of the voltage times the projected component of current to same voltage, as shown in figure 8.1. In this figure the active reading will be proportional to the product of V. I. cosϕ, that is the active power component. Shifting the voltage 900 lagging, we get a reading proportional to V. I. sinϕ, which is the reactive power component.

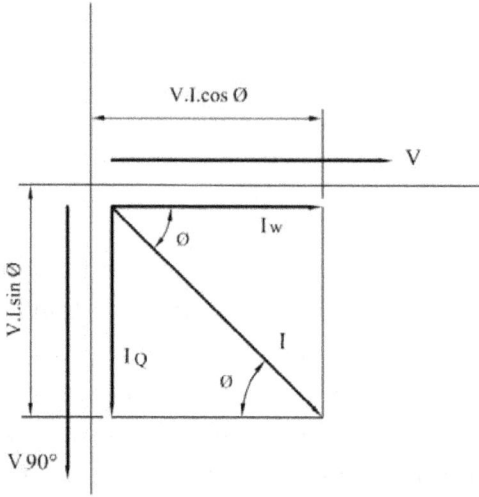

Fig 8.1 90⁰ shift of voltage vector V to read reactive power

So, connecting a voltage of the same magnitude but displaced 900 lagging to the watt-meter or watt-hour meter, the instrument will read proportional to the reactive power demanded by the system. Using varhour meters we can get the average of the demanded reactive power in a period of time. In this case it is not proper to speak about reactive energy as the reactive power doesn't do any work, therefore it should not be referred to as energy.

Referring to fig 8.1 we can see that dividing the VAR read this way by power in watt we obtain the tangent of the angle. Same thing if we use varhour dividing by watthour we get the average tangent of the angle from

which we can calculate the average power factor. The readings of varhour/watrhour will give us the average power factor during a period of time between the two readings.

Knowing the tangent of the angle, the cosine value can be found on trigonometric tables or computed using trigonometric functions.

$$\tan \phi = \text{var/watt} = \text{varhour/watthour} \qquad (8.1)$$

$$\cos^2 \phi = 1/(1 + \tan^2 \phi) \qquad (8.2)$$

$$\cos \phi = \sqrt{1/(1 + \tan^2 \phi)} \qquad (8.3)$$

$$\cos \phi = \sqrt{1/(1 + \text{kVAR/kW})^2} \qquad (8.4)$$

How can we get the 900 voltage shift? Note that the angle must be 900 but the modular value of the voltage has to be the same or has to be compensated to obtain the right reading. In the Y systems we can connect the voltage coil to VBC voltage that is placed at 900 with respect to the reference of the Y, as shown in fig 8.2

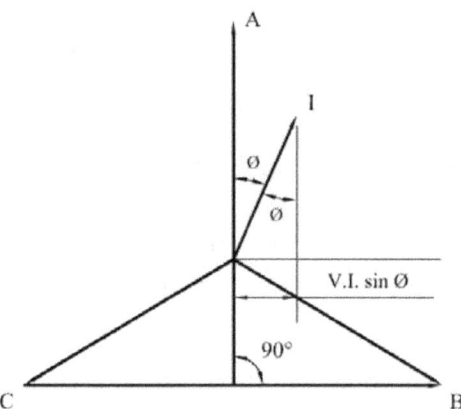

Fig 8.2 Line voltage VBC used as 900 shift to read reactive power on a Y power system.

Note that the voltage coil is being connected to a voltage that is 1.73 higher than the original phase voltage, therefore, some compensation must be made so that the reading be proportional to phase voltage. A multiplying factor must be introduced or the voltage coil of the meter must be re-wounded for line voltage. In balanced symmetrical systems this reading can be taken as a reference and the whole three-phase reading can be estimated as three times the one-phase reading.

The position of the voltage can also be shifted by connecting a resistor in series with the voltage coil (VAL) in such a way, that the voltage on the coil

will be 90^0 lagging with respect to the reference of zero angle of current in phase B, as shown in figure 8.3. In this current of phase

C must be in the same element as shifted voltage VAL. Same arrangement can be applied to the other two phases.

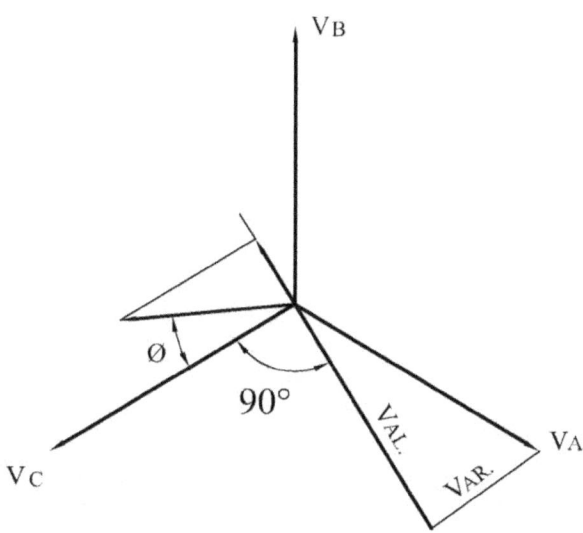

Fig 8.3 Phase shift by means of a resistor in series with voltage coil

In this case some compensation must be introduced in the meter (wattmeter or watthour meter) used for this purpose, as the voltage used to measure the reactive power will be < VA, VB and/or VC.

If the system is balanced, only one instrument can be used and find the total reactive taking three times the one-phase reading.

Sometimes phase transformers are used to get the 90^0 shift.

Figure 8.4 shows phase transformers with the right taps to connect the voltage coils of the two-element wattmeter or watthour meter to measure reactive power on a balanced Δ.

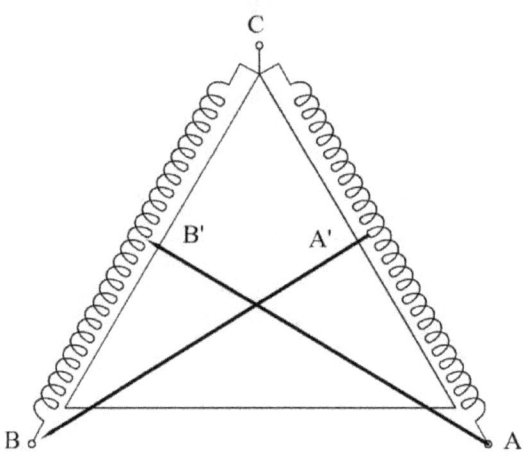

Fig 8.4 90⁰ shift by means of phase transformers

The diagram shows the voltages used to measure reactive power keeping the right polarity in the wattmeter or watt-hour meter. Polarity is very important when reading power, be it active or reactive, to obtain the correct reading. The arrowhead of the vectors indicate points of polarity

The leads of the voltage coils of the meter should separate and be connected in the right polarity in order for the meter to read correctly power or "energy" reactive.

To prove that connection in fig 8.4 reads $1.73 \cdot V \cdot I \cdot \sin\phi$ let's use a new diagram shown in fig. 8.5

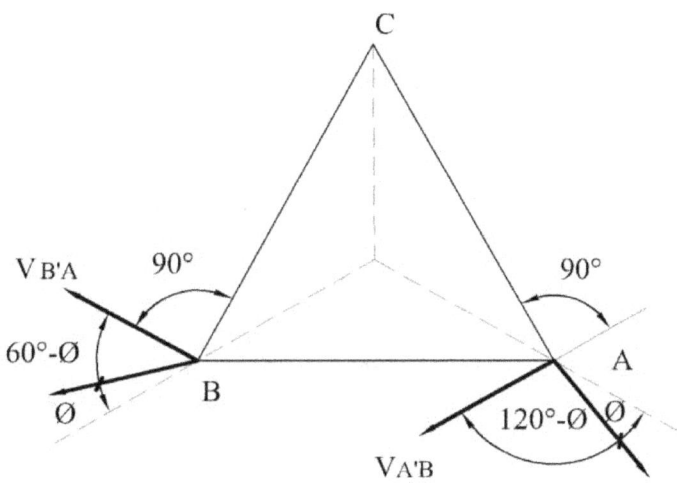

Fig 8.5 Vector relations in a triangle connection when reading reactive power

For phase C:

$V.I.\cos(60^0-\phi)$ For phase B: $V.I.\cos(120^0-\phi)$

The total power or energy must be the addition of the readings in both phases:

$$P = V.I.\cos 60°\cdot\cos\phi + V.I.\sin 60°\cdot\sin\phi$$
$$+ V.I.\cos 120°\cdot\cos\phi + V.I.\sin 120°\cdot\sin\phi \qquad (8.5)$$

But $\qquad \cos 120° = -\cos 60°$

And $\qquad \sin 120° = \sin 60°$

Therefore, $P = V.I.(\cos 60°\cdot\cos\phi + \sin 60°\cdot\sin\phi)$
$$+ V.I.(-\cos 60°\cdot\cos\phi + \sin 60°\cdot\sin\phi) \qquad (8.6)$$
$$P = 2.V.I.\sin 60°\cdot\sin\phi \qquad (8.7)$$
$$P = 2.V.I.\sqrt{3}/2\cdot\sin\phi \qquad (8.8)$$
$$P = \sqrt{3}.V.I.\sin\phi \qquad (8.9)$$

Any method we use, the main point is to use a voltage of the same magnitude but shifted 900 lagging to a wattmeter or watt-hour meter to get a reading proportional to the reactive power demanded by the customer. We have to note that the position of the voltage and/or current vector used to analyze the reading of a given connection is determined by the point of polarity of the voltage and/or current coil.

www.ingramcontent.com/pod-product-compliance
Lightning Source LLC
Chambersburg PA
CBHW071816170526
45167CB00003B/1331